Mystery Medley

Color By Number Teen

Copyright 2015

All Rights reserved. No part of this book may be reproduced or used in any way or form or by any means whether electronic or mechanical, this means that you cannot record or photocopy any material ideas or tips that are provided in this book.

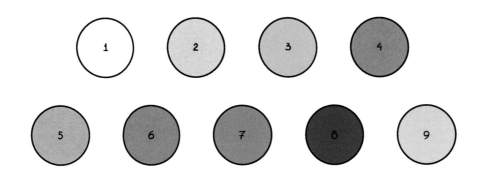
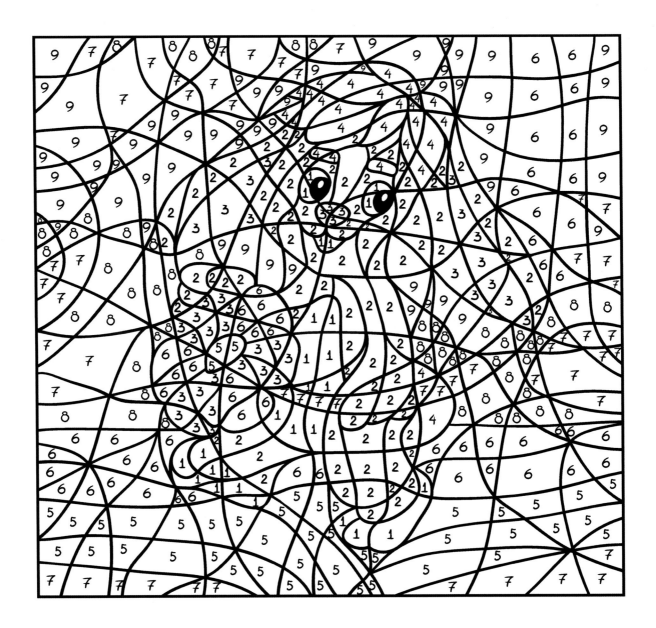

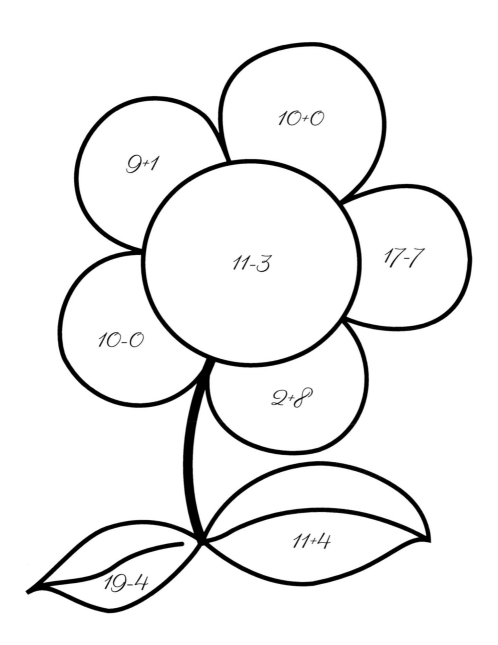

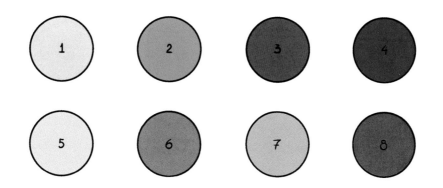

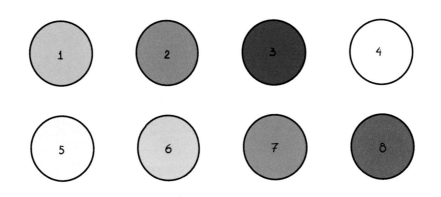

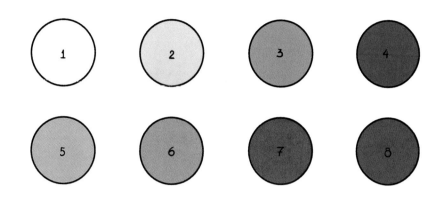

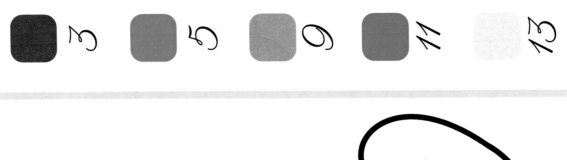
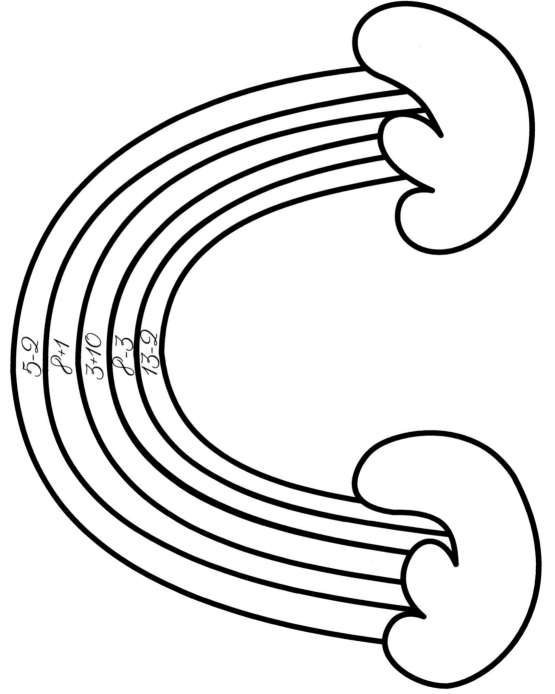

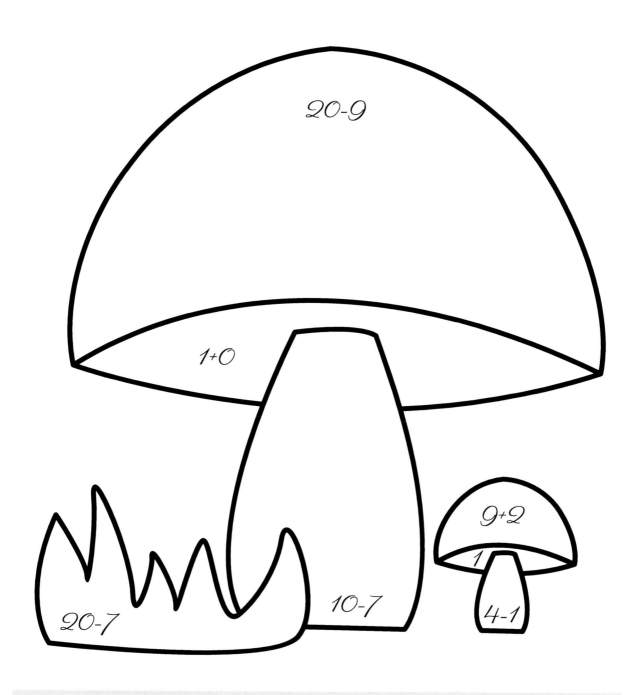

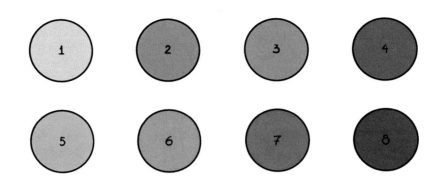

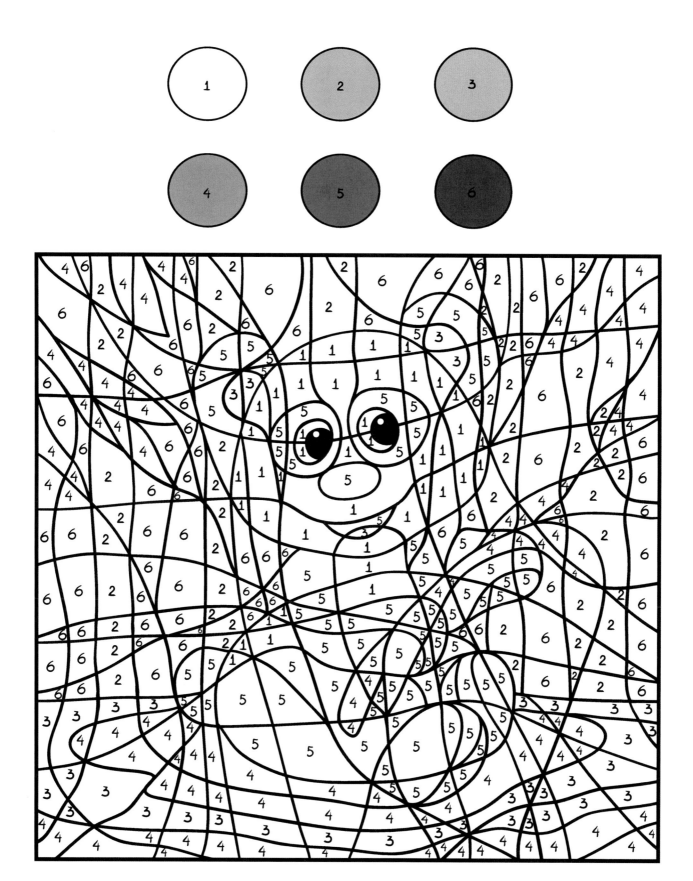

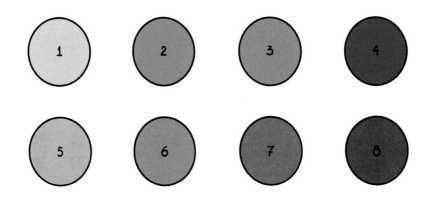

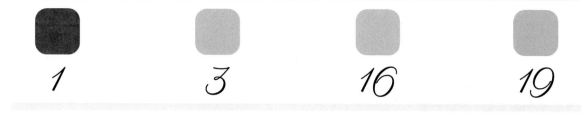

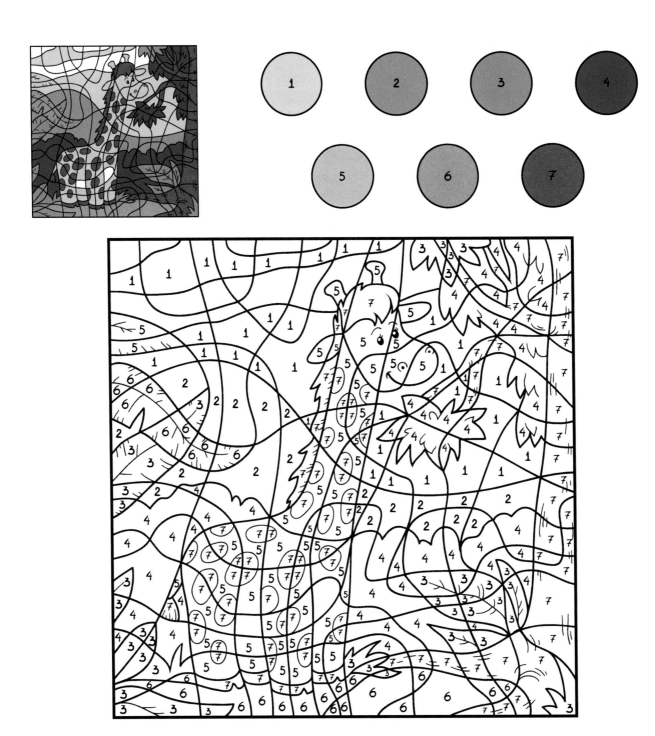

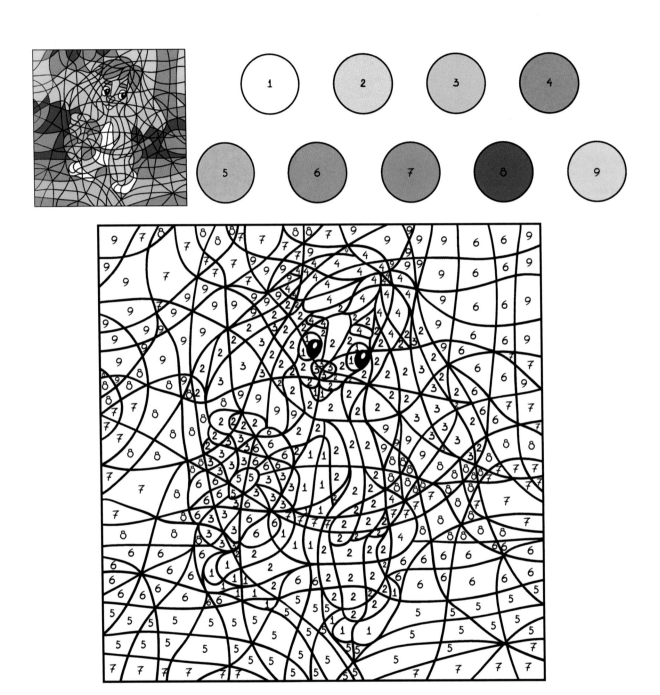

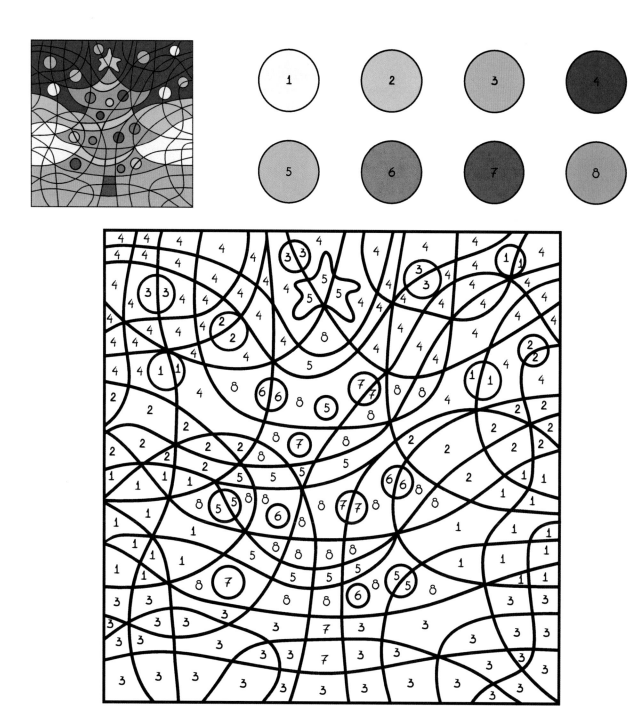

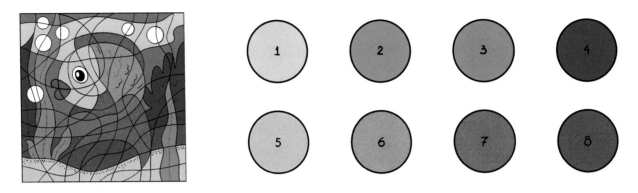

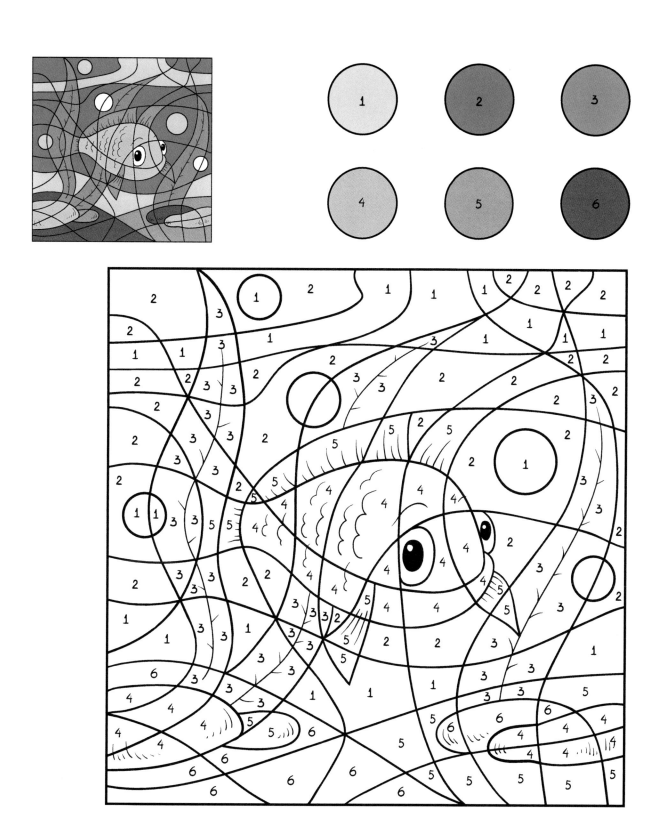

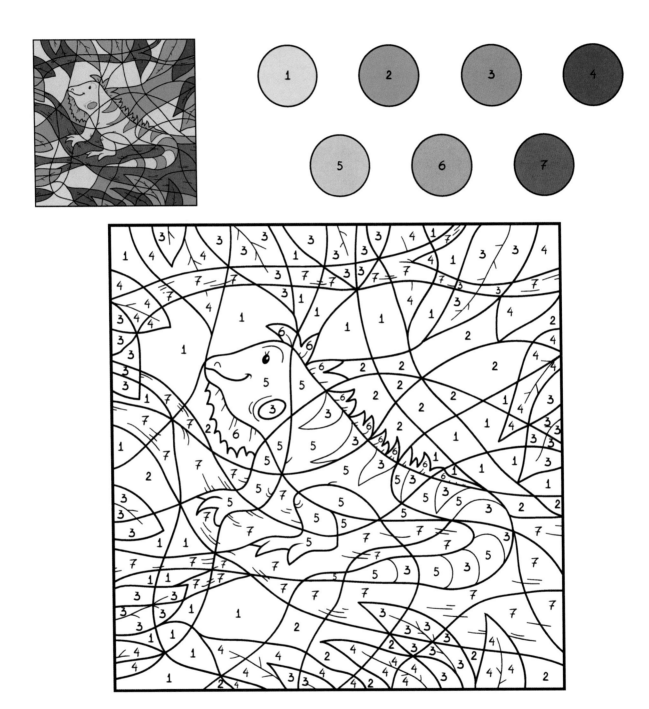

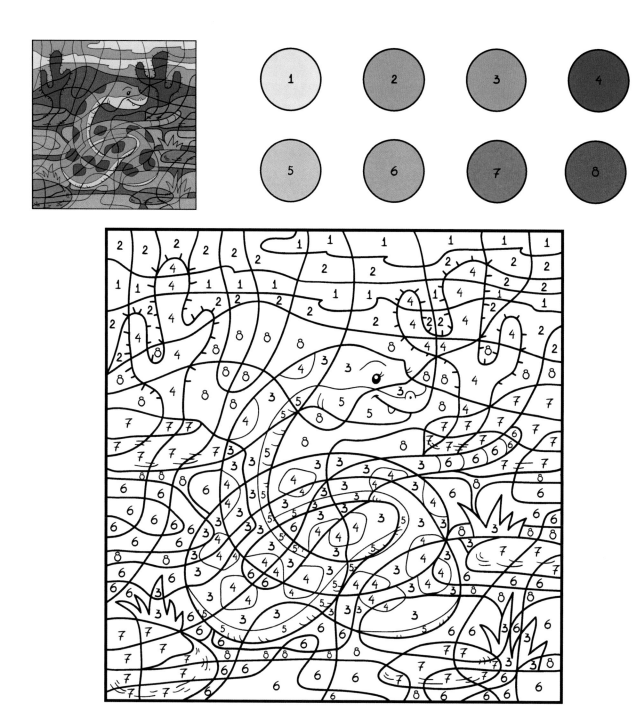

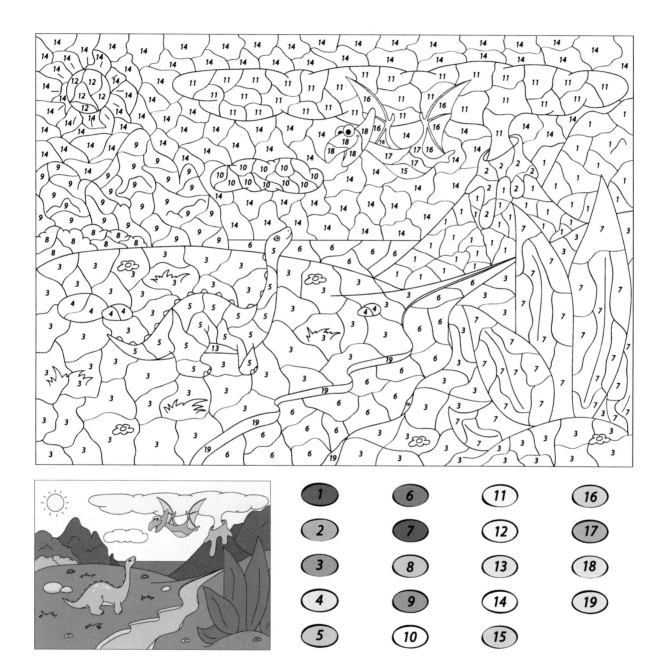

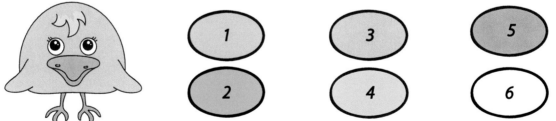

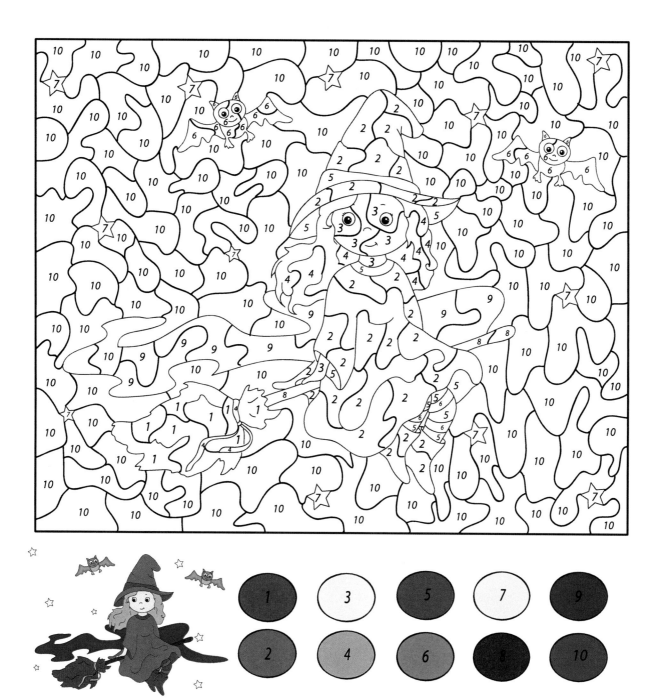

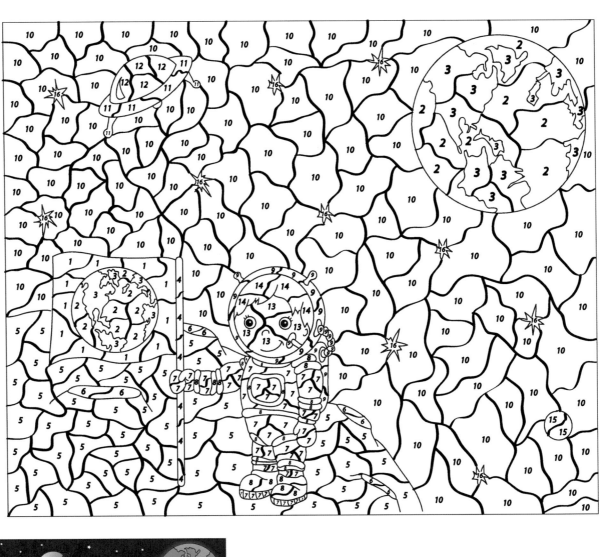

2+1 =
4-2 =
1+3 =
5-4 =
1+4 =
3+3 =
3+4 =
10-2 =

COLOR by NUMBERS

Use the color guide below to color the toy town buildings.
Color the grass green and the trees dark green. Color the sky light blue and the sun yellow. Color other things and beings as you want!

GUIDE:

1 - Red
2 - Orange
3 - Yellow
4 - Green
5 - Dark Green
6 - Light Blue
7 - Blue
8 - Purple

SAMPLE:

Paint by Number

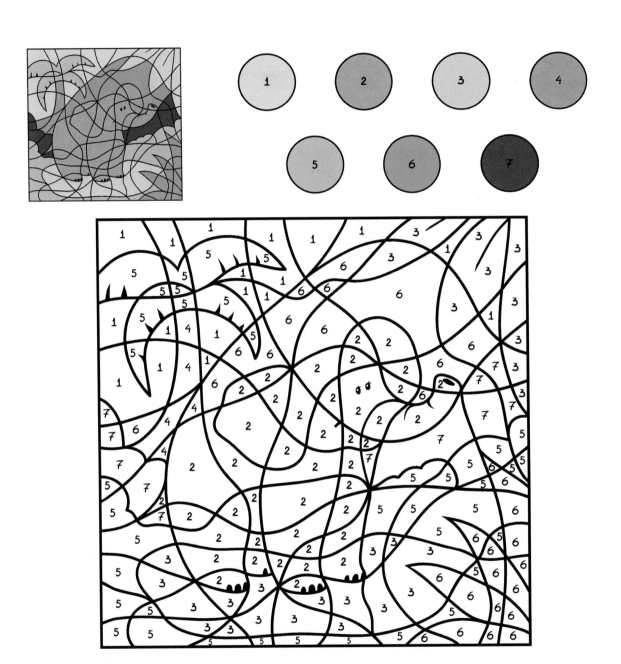

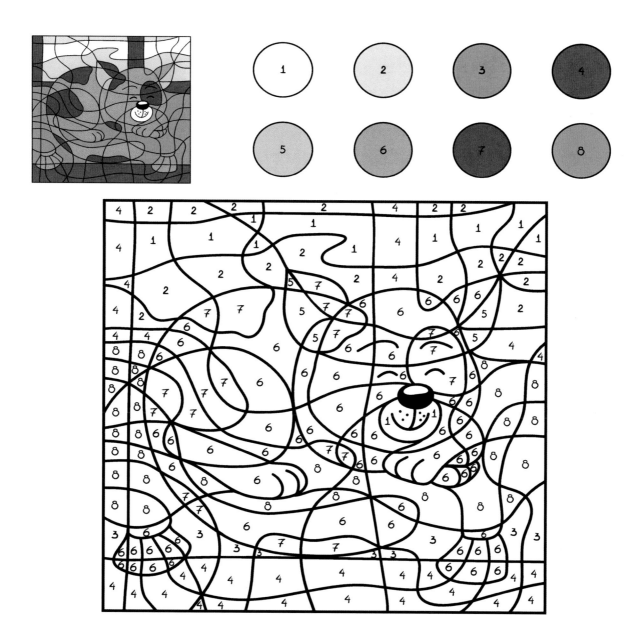

Made in United States
Orlando, FL
27 May 2023

33546943R00027